calcite,
Cumbria, UK

dioptase,
Congo

pyrite,
Illinois, USA

D0825949

stibnite,
Japan

These are crystals.

They were found within the Earth's crust and upon its surface, in the depths of mines and caves and in the mouths of volcanoes.

The crystals have not been cut or polished: this is the way they grew. The secrets of their varied shapes lie behind each crystal face.

This book can help you to discover some of the surprises hidden in your crystal world.

fluorite,
Switzerland

microcline,
Colorado, USA

hemimorphite,
Cumbria, UK

This wrapping paper, with a regular, repeating pattern on its flat surface, can help you to picture the many 3-D patterns of atoms in real crystals. A crystal is a natural and orderly state of existence for a solid substance. It is a state of internal neatness. To use and enjoy crystals it helps if you are able to picture their inner structures.

This is part of a natural silicate crystal. Magnified ten million times, this image reveals its regular pattern of atoms. Crystals are all around you – even within you – and most of the planet Earth is made of them, virtually all in the form of minerals, of which there are over 3000 known species. Crystals can also be manufactured – thus adding to this vast array of substances.

ENQUIRE WITHIN

There are thousands of different crystals, yet their structures fall into just a few basic types of pattern. Inside some crystals there are clues to their orderly atomic frameworks.

Crystal faces are natural, flat surfaces. Crystals with visible faces are relatively rare – most of the millions of crystals around you are not obviously crystalline at first sight. These may be minute particles of powder or dust, cut gems in a ring, rough grains of sand, or specially-shaped crystals and metals. Rocks and concrete contain crystals – and so does the paper in this book. Inside some crystals you can see evidence for their inner atomic neatness. Gemstones may show straight and angled colour zones or contain needles of other crystals which lie at angles set by the structure within (figs 5, 6). The cut surface of an iron meteorite can be acid-etched to reveal metal crystals (fig 3).

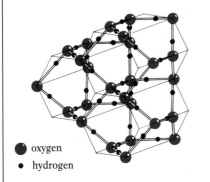

● oxygen
• hydrogen

1   Crystal structure of ice.

3   Crystal structure in an iron meteorite.

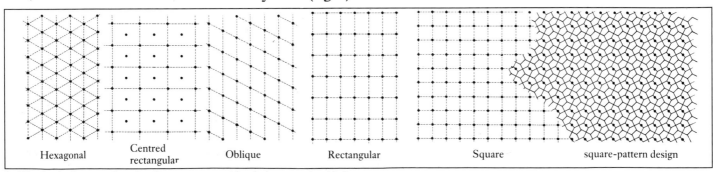

| Hexagonal | Centred rectangular | Oblique | Rectangular | Square | square-pattern design |

2   The five basic types of regular, flat pattern, one with a design example added.

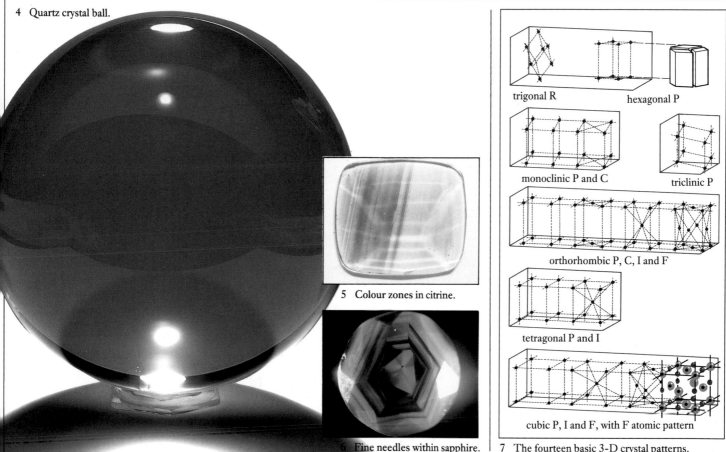

4   Quartz crystal ball.

5   Colour zones in citrine.

6   Fine needles within sapphire.

trigonal R   hexagonal P

monoclinic P and C   triclinic P

orthorhombic P, C, I and F

tetragonal P and I

cubic P, I and F, with F atomic pattern

7   The fourteen basic 3-D crystal patterns.

Although there are countless regularly repeating designs for wrapping paper, fabrics, wallpaper and floor coverings, all conform to just five basic types of two-dimensional pattern (fig 2). Crystals are natural 'designs' of atoms and groups of atoms in regularly-repeating 3-D frameworks called crystal structures. Electrons provide the links between atoms, bonding together molecular groups and whole crystals.

The precise structure adopted by atoms when they link to form a crystal depends upon which kinds of atoms are present at the time, which ones end up in the crystal structure, and the conditions (temperature, pressure, etc.) which control the way they link together.

A sapphire gem of five carats (that is one gram of sapphire crystal) is made of 10 000 million million million atoms of aluminium and oxygen which have linked together at a high temperature to form a crystal structure in a threefold repeating pattern.

There are thousands of different patterns into which atoms can crystallize, yet there are only 14 different basic types of three-dimensional pattern, the simplest pieces of which are shown in fig 7. These points represent regularly repeating patterns upon which all crystal structures are based (p.14).

PERFECT FORM & HABIT

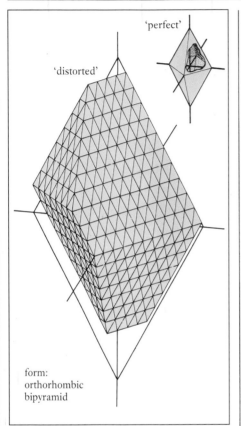

'perfect'

'distorted'

form:
orthorhombic
bipyramid

8 A single crystal form with imperfect shape – but its angles are perfect.

Crystal faces grow in a puzzling variety of shapes and sizes. To add to the confusion many crystals look misshapen – but the angles between faces are set by the pattern within.

A perfect and a misshapen crystal can have inner patterns of equal neatness. The secret lies beneath the surface.

A crystal face is a surface to which layers of atomic units have built outwards, just as the top of an office block is the last floor during the building's growth. Crystal 'floors' make up a set of parallel planes, each with the same pattern of atoms.

A crystal contains many sets of such planes (fig 8). It grows when several sets of planes build outwards (fig 11) and their varying success depends upon conditions of growth. Successful faces and their relative sizes create alternative finished shapes or *habits* (fig 10).

12 Cleavage planes seen as angled cracks in baryte.

9 Quartz crystals: which is perfect?

10 Crystal habits of pyrite.

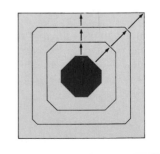

11 Slower build-up leads to larger faces.

idocrase

gypsum

baryte

axinite

beryl

13 Mineral crystals in good form.

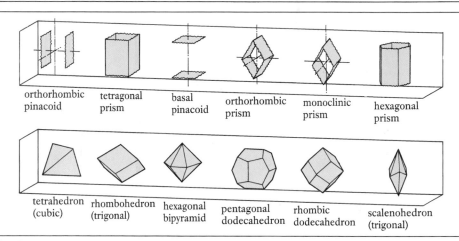

orthorhombic pinacoid | tetragonal prism | basal pinacoid | orthorhombic prism | monoclinic prism | hexagonal prism

tetrahedron (cubic) | rhombohedron (trigonal) | hexagonal bipyramid | pentagonal dodecahedron | rhombic dodecahedron | scalenohedron (trigonal)

14 Some common crystal forms: open forms are 'closed-off' by other forms in a complete crystal.

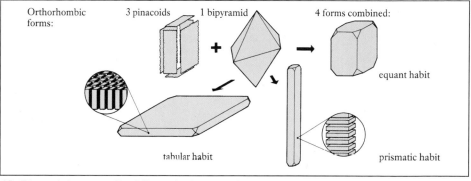

Orthorhombic forms: 3 pinacoids   1 bipyramid   4 forms combined:

equant habit

tabular habit

prismatic habit

15 Diverse habits created by differing development of crystal forms.

Sets of planes which terminate in crystal faces tend to be those whose layers are densely packed with atoms. Related sets of planes, with the same atomic pattern and spacing, may develop as a number of equivalent faces. Together, these faces are called a *form*. Fig 8 shows a single, eight-faced form. If perfect, it would have identical faces. When more than one form is present on a growing crystal, the form growing more slowly will have larger faces (fig 11). Restricted growth similarly results in a larger face, distorting the crystal shape. Face size and shape vary in real crystals but angles between corresponding faces are the same in *all* crystals of the same substance and structure (figs 8, 9), regardless of how they grew. The angles are determined by the internal structure. This relationship is also true of *cleavage*, sets of planes along which the crystal structure can be split apart (fig 12).

The form possessed by a crystal, or the relative development of different forms, determines the crystal's *habit* (figs 10, 15). A crystal's forms and habit are controlled by growth conditions and by the shape of its atomic 'building-blocks': flat blocks, like slices in a long loaf, tend to build up into a prismatic habit; long, thin blocks may stack like matches to give a tabular habit.

CRYSTAL WORLD

Planet Earth is made mostly of mineral crystals, and the bulk of these consist of oxygen atoms bound to just seven other elements. Most rocks consist of silicate crystals, where the oxygen is bound to silicon atoms.

Our 4600 million year-old Earth, nearly 13 000 km across, has a white hot interior which is slowly and continually churning around. Yet this planet is 85% crystals: it is made mostly of solid minerals (page 10). We live in a world of crystals. Life is made out of minerals, air and water – and these all come from the body of this planet. Nearly all of the rocks in the Earth's crust are crystalline. This relatively thin outer layer is three-quarters silicon and oxygen, with only six other elements making up most of the rest of its rocks. Life itself makes natural crystals (figs 17, 18).

16 Water crystals (ice): six-way angles are fixed by their internal pattern of atoms.

17 Plant crystals: tiny calcite plates from algae make up the white chalk cliffs of Dover.

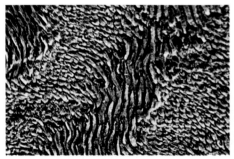

18 Animal crystals: a tough mass of apatite crystals make up your tooth enamel.

19 Mineral crystals: most pebbles are crystalline, and crystals can show up well in wet pebbles.

Oxygen makes up nearly half the weight of the Earth's crust. Yet, because oxygen atoms are very bulky, they take up more than 90% of the volume of its rocks, mud, sand and pebbles. Most surface rocks are made of minerals (p10) which are either newly-formed or recycled. Nearly all rock-forming minerals contain oxygen and their crystal structures are largely determined by their internal patterns of oxygen atoms.

Many common mineral crystals are neat arrangements of oxygen atoms, with regular small spaces between them occupied by smaller atoms such as silicon. Much of this planet is made up of well packed oxygen in crystals!

Silicon is the second most common element in the crust. It binds very strongly with oxygen to form hard polymers called *silicates*. Most of the rock-forming minerals are silicate crystals.

There are six common metallic elements whose atoms help make up the huge variety of silicate minerals in the crust. These are aluminium, iron, calcium, sodium, potassium and magnesium. Life-forms make great use of these elements, with other essentials such as carbon and phosphorus. Organic crystals of calcite (calcium carbonate, fig 17) build layers of limestone; while apatite (a calcium phosphate) is in your bones and teeth (fig 18).

MINERALS

Minerals are natural, crystalline substances. They are inorganic – not made by life-forms. Minerals make up rocks of the Earth's crust, crystallize inside it and cover its surface.

All minerals are crystalline by definition. Most rocks are made of minerals – although a few, like obsidian and pumice, are glassy and non-crystalline. The most common minerals are hard polymers, based on silicon and oxygen atoms, which crystallize in a variety of patterns called silicate structures (fig 23). Silica, pure silicon with oxygen, crystallizes as the mineral quartz (fig 21). Atoms of the common metals (p.9) occur within the huge variety of silicate minerals which make up most rocks in the crust.

    Ice and diamond are two important non-silicate minerals. Another, calcite, forms much of the world's limestone and marble – as well as stalactites.

analcime

calcite

olivine

augite

mica

quartz

feldspar

hornblende

21　Common rock-forming minerals.

20　Granite, sandstone, schist – most rocks are made of silicate minerals.

Minerals are defined as crystalline, inorganic, naturally occurring solids with definite chemical compositions. There are well over 3000 species. Minerals which make up the Earth's rocks are called *rock-forming minerals* (fig 21); most are silicates. Some kinds of rock, such as pure marble and quartz sandstone, are made up from grains of just one mineral. However, most rocks are aggregates of two or more minerals.

Minerals which crystallized from molten rock (magma) make igneous rock (fig 22); those which re-crystallized in the solid state make metamorphic rock; those which have been re-cycled, either in solution or as solid pieces, make sedimentary rock. Silicate crystals are built with strong molecular units of four oxygen atoms equally-spaced around a small silicon atom, forming a tetrahedron shape (fig 23).

Silicate minerals are made up with either separate or shared tetrahedral units within their crystal structures. The units even link up (polymerize) in molten rock, making it more viscous.

22 Feldspar, iron oxides and iron-magnesium silicate crystals (inset) make up basalt as this lava cools.

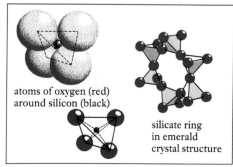

atoms of oxygen (red) around silicon (black)

silicate ring in emerald crystal structure

23 Silicon-oxygen units and a silicate structure.

GEM QUALITY CRYSTALS

Some minerals and man-made crystals are made into gemstones. To qualify, they should be hard or tough and have beautiful clarity, colour or fire. Value comes also with rarity or rare quality – and is influenced by fashion.

Many gemstones are minerals – natural crystals (p.10). These are cut and polished to display their clarity, colour or 'fire', but only good quality minerals will make worthwhile gems (fig 24). The beauty revealed by cutting and polishing is in the light which leaves the gem. Most crystals affect light in special ways, providing clues to their identification (figs 25, 26 and p.41). Gems need to be hard or tough to take a fine polish and resist wear, while fashion and rarity greatly affect the values of gems. Some minerals have all of the desirable qualities to make them ideal as gemstones (figs 27, 28).

25   Doubled image in sinhalite gem.

26   Varied colour effect as iolite gem is turned.

24   A faceted tanzanite cut from a gem-quality crystal.

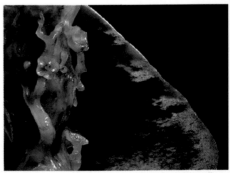

27   Jade: tough rock of interlocked crystals.

28 Gem quality crystals are faceted to reveal beauty.

Many minerals are hard silicates or oxides. These are resistant to wear and corrosion, making them suitable for continual use – and for taking a bright polish, allowing light to make the best of their colour or clarity. However, the hardest known substance is carbon crystallized as diamond, the only gem made of a single element. Substances such as jade (fig 27) have tiny interlocked crystal grains, which make them tough.

Most gemstones are minerals. However, many man-made crystals are grown with the same composition and crystal structure as minerals: these are cut as 'synthetic' gems. Other man-made materials have no natural counterpart: these are cut to make 'imitation' gems. A few organic crystalline substances appear as gems, notably pearl and coral. Some minerals, like diamond, may be rare in occurrence; others, like quartz, are common but seldom of gem quality. The rarity of gems such as emerald and star ruby lies in their interesting colour or crystalline optical effects (p 50). Gem testing makes great use of the effects of crystal structure upon light (p 40). For instance, many crystals have a varying effect on light when viewed from different directions, creating doubled images (birefringence, fig 25) and varying colour (pleochroism, fig 26).

NEATLY CLASSIFIED

Crystal structures and crystals with faces are studied and described by using their *symmetry*. This is a way of making sense of the kinds of regular, neat order that they all possess.

Many materials, such as hair and plastic, have inner structures with some sort of order: they have groups of atoms regularly arranged in fairly neat rows or layers.

Crystals have a great amount of inner order; crystal faces conform strictly to the pattern set by the orderly structure inside. Although faces can be misleading, they may reveal much about the inside. To understand and study crystals and their structure it is important to be able to describe them. The most useful thing to describe is the orderliness of crystal structure and faces, and this is found in their *symmetry* and angles: a sort of structural 'neatness'.

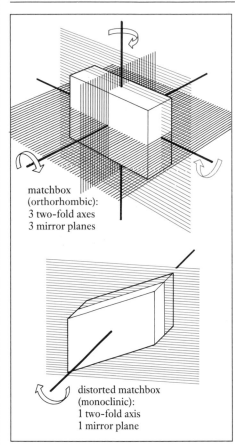

matchbox (orthorhombic):
3 two-fold axes
3 mirror planes

distorted matchbox (monoclinic):
1 two-fold axis
1 mirror plane

29 Symmetry in solid shapes.

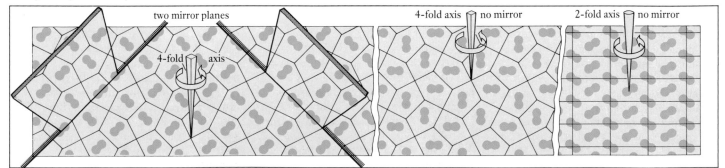

two mirror planes

4-fold axis

4-fold axis no mirror

2-fold axis no mirror

30 Symmetry describes the 'neatness' of a pattern.

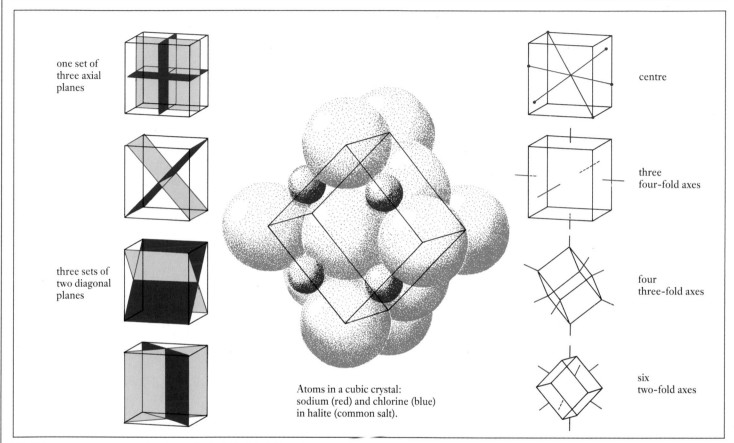

one set of
three axial
planes

centre

three sets of
two diagonal
planes

three
four-fold axes

four
three-fold axes

six
two-fold axes

Atoms in a cubic crystal:
sodium (red) and chlorine (blue)
in halite (common salt).

31   The symmetry elements of a cube; cubic building block (centre).

In using symmetry to describe and classify crystals we assume inner perfection, even though crystals never have truly perfect internal structure (page 50). You 'operate' symmetry by reflecting and rotating shapes and patterns (figs 29–31). If a pattern or shape is turned about a straight line like an axle, and the same pattern or angles appear 2, 3, 4 or 6 times in a complete turn, the line is an *axis of symmetry*.

If a mirror could be inserted, and the reflection seen as identical to the original, then the mirror is a *plane of symmetry*. Crystals which possess faces often look distorted and unsymmetrical due to accidents of growth. A *goniometer* (page 43) can reveal symmetry because angles between faces remain constant (page 6). However, flat designs and crystal structures often contain shapes which reduce overall symmetry (fig 30).

Thus the five types of flat pattern (page 4) can contain seventeen kinds of design. These account for every flat, regularly repeating design. Likewise, every crystal belongs to one of 230 kinds of 'natural design' in 3-D atomic symmetry. These are based on the fourteen 3-D lattice patterns (page 5). The 230 sets of crystal symmetry account for all of the immense variety of combinations of atoms and bonding arrangements in crystals.

CRYSTAL SYSTEMS

Crystals are grouped into *systems* according to their symmetry. You can describe the positions and angles of crystal faces with the aid of imaginary reference lines through the crystal.

The existence or absence of certain basic kinds of symmetry allows you to assign crystals to seven *crystal systems*, (fig 33). If all of their symmetry is considered, crystals group naturally into 32 different *crystal classes*. Each crystal system contains a set of symmetry classes with the same basic symmetry which is characteristic of that system.

You can describe the orientations of crystal faces by referring to sets of imaginary direction lines or axes. These reference axes, or *crystallographic axes*, act as an aid to orientation, rather like compass directions on a map. Figure 33 shows crystal forms in each of the systems with their reference axes, together with a list of their basic and characteristic axes of symmetry.

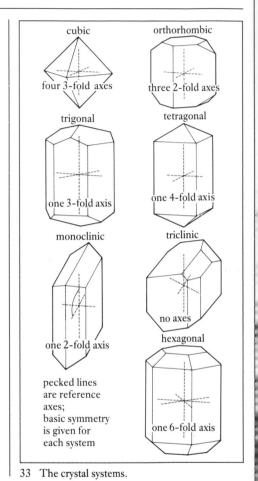

cubic — four 3-fold axes
orthorhombic — three 2-fold axes
trigonal — one 3-fold axis
tetragonal — one 4-fold axis
monoclinic — one 2-fold axis
triclinic — no axes
hexagonal — one 6-fold axis

pecked lines are reference axes; basic symmetry is given for each system

33   The crystal systems.

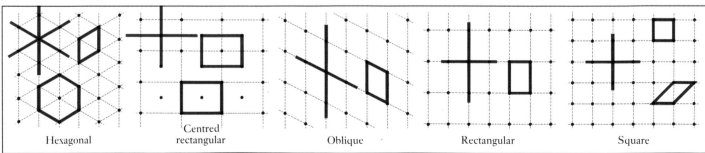

Hexagonal   Centred rectangular   Oblique   Rectangular   Square

32   Reference axes and unit cell 'building blocks' in two-dimensional patterns.

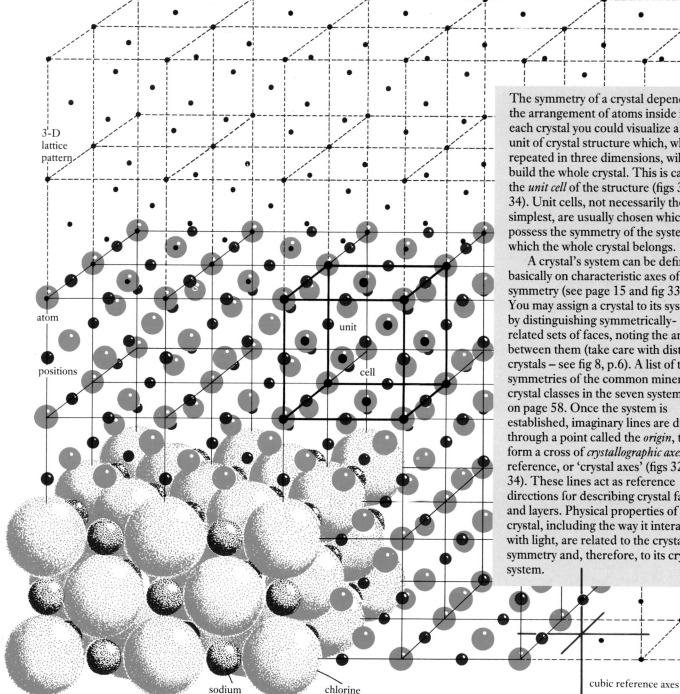

The symmetry of a crystal depends on the arrangement of atoms inside it. In each crystal you could visualize a small unit of crystal structure which, when repeated in three dimensions, will build the whole crystal. This is called the *unit cell* of the structure (figs 32, 34). Unit cells, not necessarily the simplest, are usually chosen which possess the symmetry of the system to which the whole crystal belongs.

A crystal's system can be defined basically on characteristic axes of symmetry (see page 15 and fig 33). You may assign a crystal to its system by distinguishing symmetrically-related sets of faces, noting the angles between them (take care with distorted crystals – see fig 8, p.6). A list of the symmetries of the common mineral crystal classes in the seven systems is on page 58. Once the system is established, imaginary lines are drawn through a point called the *origin*, to form a cross of *crystallographic axes* – reference, or 'crystal axes' (figs 32–34). These lines act as reference directions for describing crystal faces and layers. Physical properties of a crystal, including the way it interacts with light, are related to the crystal's symmetry and, therefore, to its crystal system.

3-D lattice pattern

atom

positions

unit cell

sodium   chlorine

cubic reference axes

34 A lattice and crystal structure in the cubic system, showing a unit cell 'building block' (centre) and 'real' atom sizes (lower left) in halite – common salt.

17

DOMESTIC CRYSTALS

In the home and out at the shops you can discover crystals – they cool your drinks, grow in your kettle, stir the tea, cover the walls and relieve your headache.

Crystalline substances help smooth your lives – the spark of a gas igniter comes from a special crystal; orderly atoms in polarizing sunglasses 'tidy up' light.

You see prepared crystalline materials when you look at facing stone on shops and offices, at kerbstones and lamp-posts, vitamin pills and pain killers; metal spoons and buckets; paint, paper and face powder (but not crystal glass!). Salt and sugar are recrystallized for your food, while new crystals grow in setting concrete.

Most gems in your jewellery – and many of the dust particles you breathe – are made of crystals. Crystals grow right next to you – the 'fur' in the kettle and ice from the freezer.

35   Agate snuff bottle, Chinese, 19th century.

36   Feldspar crystals in facing stone: larvikite in a London shop front.

37   Crystals at home include 'fur' in your kettle (magnified 850 times).

WILD CRYSTALS

If you are at the seaside, on a mountain holiday, travelling across country, or maybe walking across the desert, you just can't help gazing at crystals – or treading on them.

Tiny mineral grains, which are crystalline particles, make up whole mountains. Their peaks are coated in crystalline ice. You can watch as sand and mud, also crystalline, are swept off to make layers of new rock.

38 Large crystalline masses.

You can take home seaside pebbles for a closer search under a lens for the crystal shapes of mineral grains. On salt lakes you see dissolved minerals crystallizing out of warm water. In museums you may discover all sorts of crystals – maybe your local minerals, rocks and ores.

Outside this tiny planet Earth there is a vast cosmic stock of crystalline dust and lumps of rock and metal. A few of the larger pieces reach the Earth's surface; these are called meteorites.

40   Exotic crystals: a stone meteorite, Barwell, Leicestershire.

39   Lake salt.

41   Trapped crystals!

42   Pebbles of seaside crystals.

MICRO GALLERY

We use light or electrons in microscopes to give magnified images of crystals, their forms, markings and cleavages. We identify minute mineral grains by prying with light.

Microscopes and hand-lenses allow you to pry into the depths of gemstones (fig 48), providing you with clues for gem identification and origins, and revealing fakes.

It may be too difficult to pump oil from sandstone reservoirs if the rock's pores are blocked by crystallized minerals. A scanning electron microscope reveals the scale of the problem (fig 51).

Slices of rock three hundredths of a millimetre thick are studied between polarizers (figs 43 to 47) in a *petrological microscope* (see page 41).

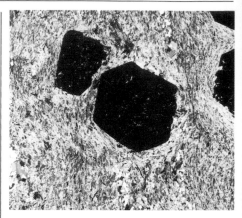

43 Garnets in schist show dark amid mica flakes.

45 Granite magnified in polarized light.

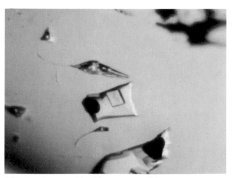

48 A salt crystal inside an emerald.

49 Metal crystals in close-up: brass.

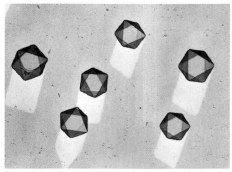

50 Silver bromide crystals in photographic film.

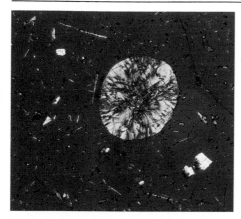

44 Crystals grow within volcanic glass.

46 Moon rock in polarized light, Apollo 11.

47 Agate: layers of minute quartz particles; Polarized light.

51 Quartz sandstone with blocked pores.

52 Granulated sugar.

53 Vitamin C crystals.

LESS THAN PERFECT?

'Perfect' crystals are found only as text book diagrams and models. Real crystals are never truly perfect, yet their imperfections may be beautiful and fascinating.

Some museums contain beautiful crystals, notably in mineral collections. Yet, of all the crystals in the world, relatively few have faces; even when crystals do have faces, they seldom conform to the ideal forms seen in text book diagrams and crystal models. Beautiful and high quality specimens (fig 56) normally have some surface markings and imperfections in their face shapes.

    Most of the crystals you can handle are in some way 'imperfect'. Such imperfection is in the build-up of planes of atoms and in the resulting face details, not necessarily in beauty, value or interest. Indeed, structural imperfection within crystals is of great and often vital interest (p.50).

56 Quartz crystal group.

57 Needles of mesolite.

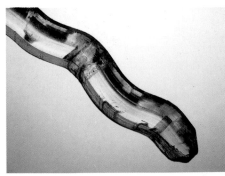

54 **An imperfect gypsum crystal.**

55 Twin layers in feldspar.

58 Curved faces in diamond.

No crystal has perfect structure or form. Alternating attempts by different sets of planes to form their own face lead to striations across faces. Different rates of growth on different parts of a face produce effects such as 'hoppers', curves, needles, dendrites or frost patterns. 'Flip-over' in crystal structure creates *twins* (figs 55, 59, 62) during growth or under stress or through change in pressure or temperature.

59 Twinned cerussite.

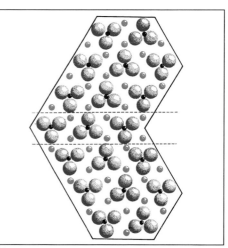

62 A twinned crystal structure.

60 Tree-like manganese oxides 'dendrites'.

61 'Skeletal' cerussite.

63 'Hopper' growth in salt.

This is a picture of crystals growing. Under the microscope you can see them grow or *crystallize* from solution at the rate of around a million billion atoms each minute.

Atoms, or groups of atoms, add to their outer surfaces, building up the orderly crystal structure.

At that growth rate it is not surprising that crystal structures and surfaces are never truly perfect.

Spiral shapes on some crystals reveal the way in which swathes of atoms can build up a crystal structure. They do this only if there are suitable defects in the crystal. In fact, many crystals grow more quickly if they are *imperfect*.

Flat surfaces or *faces* are relatively rare – growing crystals may get in one another's way; they finish up as a random mass of crystals grown together.

BIRTH & SURVIVAL

Crystals grow and survive in all conditions – in arctic seas, deserts and furnaces, around the stars in space, inside plants and animals (including you) and deep inside the Earth.

Whether a crystal grows in the Earth or in a factory, it does so because natural forces are acting in the right environment for that particular crystal to survive. For instance, diamond may grow naturally or artificially from a supply of carbon atoms under great pressure and heat in the right chemical environment.

If any crystal is moved away from its birthplace, new conditions might destroy or alter it in some way. Diamond survives well because it is very hard, but it burns if heated in air. Any one set of atoms can form alternative crystal structures, depending on conditions in which it crystallizes – carbon may form diamond *or* graphite (fig 66).

65   Diamond from 250 km underground.

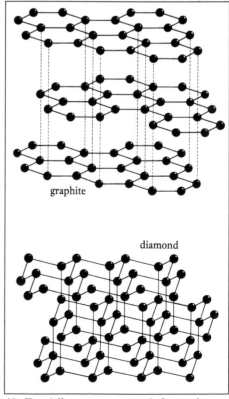

66   Two different structures made from carbon.

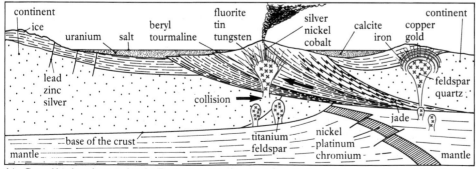

64   Crystal birth and survival in the Earth, showing where metals and other substances gather as minerals.

67 Diamond survives well on the Earth's surface.

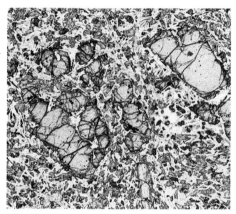

68 Corroded crystals in lava.

69 Volcanic ash often contains crystals.

The Earth is white hot inside, over 1500 degrees Celsius below 200 km down; and the pressure is enormous. The solid, crystalline rock making up the Earth's mantle slowly circulates; continents crush together (fig 64). In certain places rock is melted and this *magma* later crystallizes as it cools slowly beneath the surface. Some magma reaches the surface, creating volcanoes, scattering dust (fig 69) or maybe pouring out as lava (fig 22, p.11).

Crystals of the mineral olivine grow in very hot magma. Olivine is not stable in the chemical environment of magma at lower temperatures and becomes corroded as it cools (fig 68). Large, gem quality olivine crystals are cut to make the gemstone *peridot*. Feldspar crystals may grow in molten rock which is then erupted from a volcano; the crystals in a fine groundmass give lava a *porphyritic* texture (fig 79, p.33).

Crystals grow in space, around brand-new, dusty stars and old-aged stars which puff out their 'star dust'. Much space dust is made of very tiny particles, but it includes some familiar minerals. The dust gradually accumulates to form new stars and planets. Meteorites consist of crystalline rock and metal; some are chunks of planets, moons and asteroids. Crystals also grow inside you, in hard bones and teeth (fig 18, p.8).

HOW CRYSTALS GROW

Crystals grow where atoms gather in an orderly manner. Many crystals grow from atoms in liquids – watery solutions and molten rock. Others grow from vapour and in solid rock.

In our everyday experience, things seem to fall apart and disperse easily and are hard to put back into order. Yet when atoms, or groups of atoms, bond together they usually manage to 'shake down' into an orderly pattern. Once a small number of atoms exist in a minute, regular group – a speck of a few hundred atoms perhaps – more atoms may continue to bond onto it. Tiny solid grains make ideal starting points for new crystals.

Defects aid the growing process (figs 72, 73) and the orderly structure grows outwards until the supply of atoms stops. A state of solid *disorder* is called 'amorphous'. Glass is amorphous (fig 127, p.50), so 'crystal glass' is non-crystalline!

70 Menthol crystals grown from vapour.

71 Alum crystals grown from water solution.

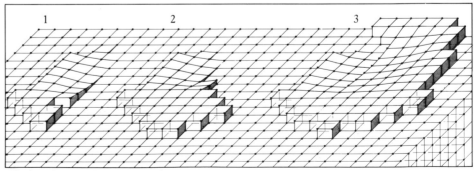

72 Spiral growth from a screw defect.

73 Spiral growth on a crystal face.

Grow crystals from solution (page 32) by dissolving potash alum in warm water. Stir until no more dissolves; pour into a jar. Cool, and suspend a single alum crystal 'seed' on nylon thread from a hook (fig 74a). Leave the jar open for the solution to *evaporate*. Excess alum adds to the seed over a few days.

Alternatively, having poured the saturated solution, *seal* the jar and cool to a steady temperature. Any alum which can not remain in solution at the lower temperature adds to the seed crystal (fig 74b). Evaporate some solution in a dish to provide seed crystals. You can also grow crystals of Glauber's salt, common salt, chrome alum, copper sulphate, citric acid and borax.

To grow crystals from molten liquid (p.32) and vapour (p.34) place crystals of menthol in a small, clear jar. Hold the outside of the jar in fairly hot water until the crystals melt (at 44°C) then run the liquid around the inside of the jar. Watch the liquid crystallize. Keep a sealed jar of menthol (figs 70, 74c) in a warm place and see crystal needles grow over the weeks and years.

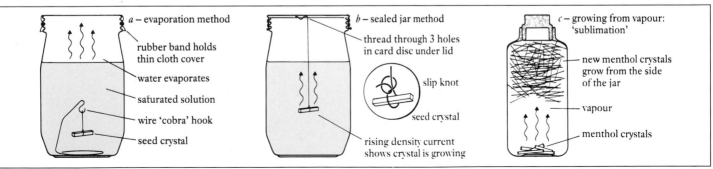

a – evaporation method
rubber band holds thin cloth cover
water evaporates
saturated solution
wire 'cobra' hook
seed crystal

b – sealed jar method
thread through 3 holes in card disc under lid
slip knot
seed crystal
rising density current shows crystal is growing

c – growing from vapour: 'sublimation'
new menthol crystals grow from the side of the jar
vapour
menthol crystals

74 Growing crystals – a, b: in solution; c: from vapour.

GROWING IN LIQUID

Most natural crystals grow from liquid solutions. Water or molten rock carry atomic building blocks which may bond together – crystallize – when conditions change.

Some solid substances dissolve in liquids. Usually the same solid can reappear as it separates from the 'carrier' liquid – it crystallizes out. This can happen, for instance, when the temperature changes or the liquid becomes 'full', or *saturated* with the dissolved substance. The carrier liquid – the *solvent* – may be water, molten salt, molten metal, molten rock (magma) or organic liquid solvent.

Molten rock is a complex mixture of atoms and molecules. While magma cools, some of its atoms group together in an orderly fashion: they crystallize from the molten liquid. Pure molten substances crystallize directly as they cool to *freezing point*, for instance water at 0 degrees, or sapphire at 2035 degrees Celsius.

76   Crystals of synthetic ruby and sapphire.

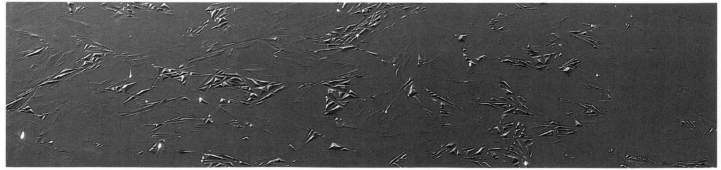

75   Ice: water crystals growing in a pond.

77 Crystals of saccharin grown from solution.

78 Synthetic quartz grown from hot water solution.

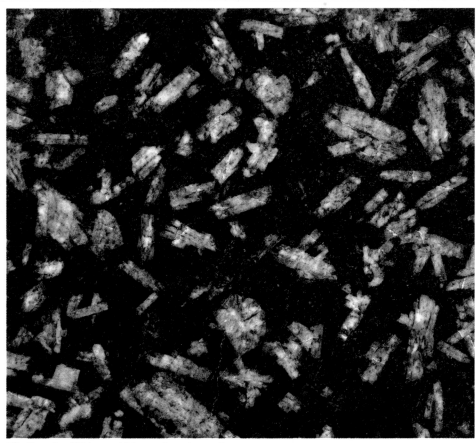

79 Crystals of feldspar which grew in molten rock.

Solutions are formed when atoms, ions (charged atoms) and small groups of atoms are dispersed – *dissolved* – within a liquid 'carrier', or *solvent*. The liquid will carry dissolved substance up to a certain limit at which it becomes *saturated* with it. Then the liquid solution may begin to 'dump' its load of dissolved substance – it crystallizes. This happens, for instance, if the temperature of a saturated solution drops, and fewer atomic particles can be carried in that solution at the lower temperatures. They have to appear as groups of solid particles again, and do so by crystallizing. Beyond saturation point, more substance may exist in solution to make it *supersaturated*. For crystals to grow, therefore, a triggering event may be needed, such as an influx of crystalline 'seed' particles to act as nuclei. Crystals grow from these until the solution is just saturated.

Typical minerals which grow from watery solutions include calcite (including stalactites), fluorite, lead and zinc sulphides, salt, silver and gold.

Minerals crystallizing from molten rocks to form *igneous rock* include olivine, quartz, mica, hornblende and feldspar (fig 79). Hot, watery, silica-rich fluids crystallize as *pegmatites*, sometimes with large crystals including topaz, beryl and zircon.

GROWING IN SOLID & VAPOUR

Crystals grow within solid materials and also directly from vapour and gases. Crystals in a solid mass can each grow larger or change in shape or internal structure as conditions change.

Atoms in many solid materials tend to 'settle down' into a tidier arrangement as time goes by. For instance glass and toffee can lose their state of non-crystalline disorder and re-form as a cloudy mass of tiny crystals, each with its regular pattern of atoms (p.3). Existing crystals may grow larger when their inner structures are rearranged, often when they are warmed or under pressure. Under extreme conditions, new crystals can form when the atoms of the original crystals are completely rearranged without any melting (figs 81, 82). Some substances will crystallize from vapour without any intermediate liquid being formed (figs 80, 83).

80 Ice crystals grow as frost directly from water vapour.

81 Crystals reorganized: shale (left) becomes garnet-studded mica-schist.

82 Crystals reorganized: limestone (left) becomes marble.

83   Yellow sulphur crystals grow around a volcanic fumarole.

High pressure can recrystallize solid rock, making new crystals out of old. The mineral grains making up the rock may be re-shaped, enlarged or converted into different minerals. At extreme pressures one kind of crystal structure is converted into another, denser structure. This *phase transformation* accounts for different density layers within the Earth and other planets; it can also be achieved artificially (p.55, fig 143).

Steel is 'annealed' to make larger crystals. This also happens when limestone, with its minute calcite crystals, is naturally heated and/or pressurised in the crust to form marble (fig 82). This rearranged rock has larger, sugary calcite crystals. Shale crushed under newly-rising mountains has rearranged atoms, seen as new minerals such as garnet (fig 81). Rocks recrystallized by such atomic rearrangement are called *metamorphic* rocks.

Given time, substances such as ice-cream, stalactites and lime-mud will recrystallize. Sudden transformations can occur under extreme pressures, such as shock from collisions: dense forms of silica occur around some large meteorite impact craters.

Certain crystals grow directly from gas or vapour by capturing atoms and molecules which do not condense to a liquid state (figs 80, 83).

EVIDENCE OF CRYSTAL GROWTH

Many crystals bear evidence of their origins. Zones of colour or impurity mark stages in their growth. Surface defects and trapped bubbles or crystals may reveal their histories.

Crystals frequently grow in successive stages rather than in one steady episode. On close inspection they may reveal these stages in a series of colour zones (fig 89) or internal ghost-like crystal shapes (fig 84) where a dusting of minute crystals or bubbles has coated the crystal between episodes of growth (see pp.4, 5).

Markings on crystal faces reveal the way in which crystal patterns build up during growth. Figs 87 and 88 show growth marks at different scales. You will see such marks on most crystals, especially if you view them through a hand lens.

Inside many crystals you will see *inclusions*, some of which formed during the growth of their host crystal. Inclusions are usually tiny crystals (fig 86), cavities or bubbles. Sometimes, cavities contain gas or liquid (p.51, fig 132).

84 Growth stages in quartz show as 'ghosts'.

85 Solution pits on aquamarine – their shapes are controlled by the crystal pattern.

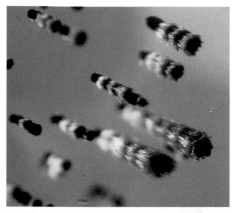

86   Tiny goethite crystal tufts within amethyst.

87   Growth architecture on a galena crystal face.

88   Diamond growth marks under the microscope.

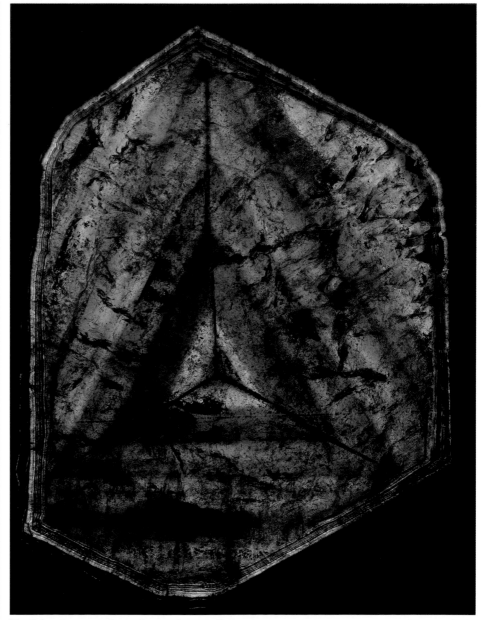

89   Colour zones reveal growth stages in tourmaline.

This photograph reveals the interaction between crystals and light waves. A very thin slice of rock, basalt lava, is seen in polarized light. Its mineral crystals rearrange light vibrations, which then re-combine to create light and colour effects.

Orderly atomic patterns in crystals also affect the way in which crystals interact with radiation and electricity, and are affected by abrasive wear, heat and pressure.

These effects can help you to identify and enjoy the enormous diversity of crystals.

This picture of symmetrical spots is a record of the interaction of X-rays with the pattern of atoms in a crystal of beryl. Most of what we know about the atomic structures of crystals comes from X-ray studies.

Detailed study of crystals allows us to control their growth and to produce and 'design' crystals to ever greater advantage – to discover how materials behave and fail, how the planets and their rocks and minerals were formed and how these can best be exploited.

CRYSTALS & RAYS

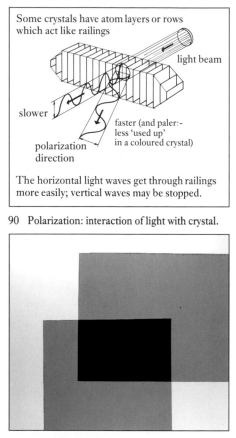

Some crystals have atom layers or rows which act like railings

light beam

slower

faster (and paler:-
less 'used up'
in a coloured crystal)

polarization
direction

The horizontal light waves get through railings more easily; vertical waves may be stopped.

90 Polarization: interaction of light with crystal.

91 Two crystalline polarizers, 'crossed'.

Regular patterns of atoms making up crystals reorganize light and other kinds of radiation. The effects can help you to identify crystals; the secrets of crystal structures are revealed.

When you 'see' a substance, it is because it is doing something to light rays before they reach you. Crystals interact with light and other radiation – such as X-rays or ultra-violet light – in an organized fashion. Light entering most types of crystal is made to conform to their organized pattern of atoms (figs 90, 91, 95, 96): the light is 'tidied up', and can then be used to help identify the crystal. This *polarized light* can be 'extracted' and used to pry into other substances.

X-rays open up secrets of structure in detail. Interaction of X-rays and powdered crystals is used in routine identification: lines (fig 94) record rays scattered by regular patterns of atoms in crystals.

92 Polarized light rearranged.

93 Sandstone slice between crossed polarizers.

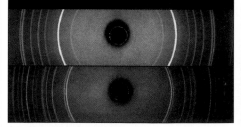

Two different crystal structures made of the same atoms: calcite (top) and aragonite, both calcium carbonate minerals.

94 X-ray powder photographs.

95 Clear calcite, Iceland spar, gives doubled image.

Calcite crystal, trigonal calcium carbonate, ideally demonstrates the way in which crystal structure and light interact. Light energy is constrained to vibrate in two planes at right angles (fig 96) by the electronic effect of the layers of atomic bonding. Each plane is of *polarized light*. Cubic crystals do not have this effect; nor do non-crystalline substances. A polarizer (figs 90, 91) absorbs one of these planes. The remaining plane of polarized light can be used as a 'light-knife' to pry into other crystals (page 42). Crystalline substances rearrange light to suit their own structural patterns; a second polarizer re-combines the light to reveal crystal effects (fig 92). Different effects are seen in different directions. Rock slices, a standard three-hundredths of a millimetre thick, are viewed under a microscope in polarized light (fig 93), aiding crystal, mineral or rock identification. X-rays are 'tidied up' in a different way. Their shorter wavelengths are scattered, diffracted by electrons in regular atomic bonds. Photo patterns are the key to crystal structures (fig 94).

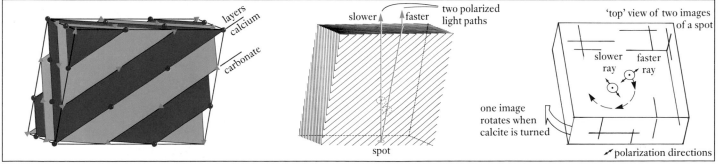

96 Clear calcite reveals the interaction of a crystal atomic pattern with light, causing 'double refraction'.

HOW TO LOOK AT CRYSTALS

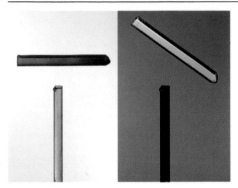

97 Polarized light through tourmaline crystal reveals *dichroism*: variable colour (left) and *extinction*: light cut-out by second, 'crossed' polarizer.

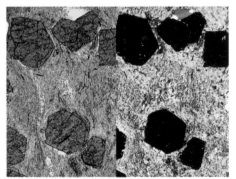

98 Garnet, quartz and mica in schist rock slice: in polarized light (left) and in 'crossed polarizers'.

A close look at most solid substances reveals their crystalline nature. You can use a lens to see order and symmetry in surface and inner details, and you can test for optical effects.

Close inspection of crystalline materials can reveal order and symmetry. Sometimes, markings on crystal faces are prominent and their symmetry can help you identify a crystal or assign it to a system or symmetry class. Trigons (fig 102), for example, are growth markings which result from different planes in the atomic pattern building up in successive steps; they reveal three-fold symmetry. You can view colour changes in different directions through some crystalline objects such as gemstones with a simple *dichroscope* made of polarizers or calcite (fig 106). Test crystals between crossed polarizers (p.40 and fig 97) with a simple *polariscope* (fig 107), turning the crystal around.

102 'Trigon' growth marks on a ruby crystal.

99 Cleavage seen in powdered fluorite crystal.

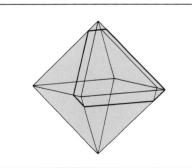

100 Cleavage in fluorite.

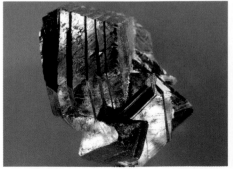

101 Twin planes in the surface of zinc blende.

Make a goniometer (fig 104) to measure face angles to check symmetry and aid identification. The internal control of cleavage – breakage along crystal planes – is apparent when crystals are powdered and magnified (figs 99, 100). Twinned crystals (p.25) often possess 'valleys' (fig 101). Since crystals usually have definite chemical compositions, their physical and optical properties are quite constant. This means that properties like *refractive index* – 'optical density' – and double refraction (p.41, fig 95) can be used to help identify many crystals, minerals and gems (fig 105).

104 Goniometer for face angles.

103 Using a high-power lens: keep it to your eye, with specimen in bright light.

105 Measuring the refractive index of a gemstone.

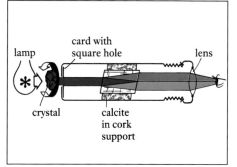

lamp

card with square hole

lens

crystal

calcite in cork support

106 Make your own dichroscope in a tube.

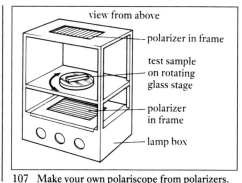

view from above

polarizer in frame

test sample on rotating glass stage

polarizer in frame

lamp box

107 Make your own polariscope from polarizers.

COLLECT YOUR OWN CRYSTALS

108 Mineral collecting.

109 A miner's 'spar box'.

From pebbles to diamonds, from microscopic specks to giant rocks and sculptures, you can build your own collection of crystals and crystalline materials for study and enjoyment.

Your crystal collection may reflect your particular interest in the worlds of natural or man-made materials.

Natural inorganic crystals – minerals – make ideal collections (fig 111) and may follow a chemical or geographical theme or depend on colour, form or rarity.

Pebbles reveal crystals well (p.9, fig 19), make useful rock collections and remind you of pleasant, distant places.

Microscope owners can get down to detailed perfection and fascination with micromounted minerals (fig 113) and thin-section rock slices (pp.40, 42).

Gem collections (fig 112) may be set in jewellery, or perhaps stored in private collections for study or investment.

110 Pebble collecting in Scotland.

111 Part of a large mineral collection.

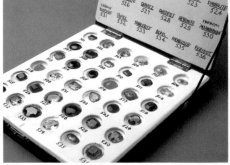

112 A gem collection.

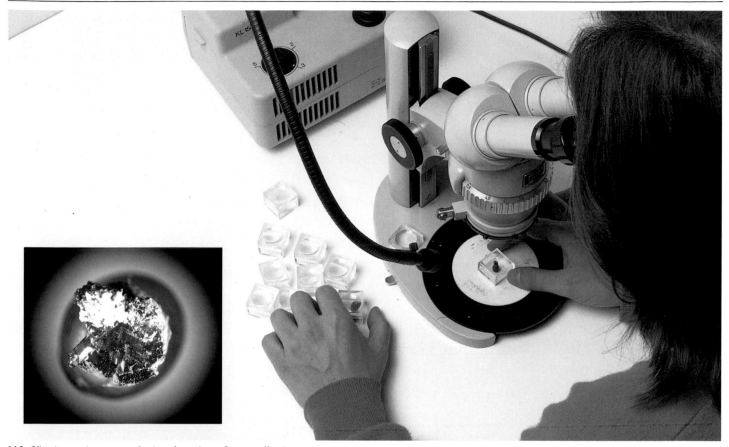

113 Viewing a micromounted mineral specimen from a collection.

Micromounts are permanently mounted mineral specimens, varying from pin-head size to twenty millimetres square. With their small size they can be of almost perfect crystal form, often with exquisite colour.

Rare minerals and unusual groups come in affordable sizes, allowing enjoyment of their beauty and yet occupying very little space.

A small shoe box will hold a collection of 200 mounted specimens.

For help and inspiration in collecting, visit museums and libraries. Museum staff and reference collections may help you identify specimens and objects; some museum libraries hold useful journals, text and picture books on geology, minerals, gems, fieldwork, lapidary and chemistry. Public libraries may hold lists of clubs, societies and study courses, while gem and mineral shops and shows provide a range of crystals.

Although your primary intention may not be to collect crystals, you achieve just that if you ever feel inclined to put aside or preserve baby teeth (see page 8), or gall and kidney stones. A collection of natural, organically-grown substances containing crystals might include these, together with coral, pearls, fossil and modern shells and bone and most limestone. Sugar, menthol and citric acid crystals are also organic in origin.

CULTIVATED CRYSTALS

We rely upon artificially cultivated crystals, from de-icers to steel-strengtheners. Their growth develops under natural forces which can be controlled in various ways.

Crystalline substances can be prepared with very high purity. Few types of natural crystal have technological uses; many substances which are not found naturally are, however, crystallized for particular purposes.

Crystals are also carefully re-formed in substances which rely on shape, size and type of crystals in them to impart required properties. Metals and alloys rely on such treatment for strength, and ice-cream for its texture.

Artificial gems and rods for lasers (fig 117) and 'chips' are grown by high-temperature 'freezing'. Diamond grit is grown at high pressure; quartz is grown from hot water solution. Crystalline coatings for lenses and optical devices are grown from vapour or solution.

115 Cultivated habits of salt for different uses.

114 Carborundum crystals and abrasive grit.

116 Carborundum crystals in close-up.

117 Synthetic crystals: yttrium aluminium garnet (YAG) for lasers.

118 Synthetic crystal growing.

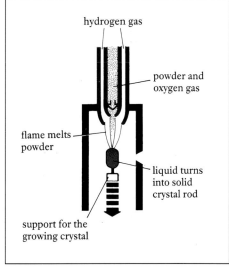

hydrogen gas

powder and oxygen gas

flame melts powder

liquid turns into solid crystal rod

support for the growing crystal

119 Verneuil flame-fusion furnace.

Flame-fusion 'freezing' in the heat of an oxygen-hydrogen flame (fig 119) is a rapid method for producing crystals of high enough quality to be cut as gems, and for bearings in precision engineering. Powder is sifted into the flame to melt onto a holder which is slowly lowered. The melt freezes to form a solid crystal such as ruby or sapphire. More perfect crystals are grown by pulling a crystal rod slowly from molten liquid (figs 117, 118).

Synthetic quartz crystals (fig 78, p. 33) are made from hot, pressurized water solution, using quartz plates as growth 'seeds'. Synthetic emerald is slowly crystallized from solution in a molten 'flux' of metal oxides. Diamond is synthesized under very high pressure and temperature by converting carbon to a cubic structure. Carborundum (figs 114, 116) is made with silica rock, carbon, sawdust and salt in electric furnaces.

Study of natural crystals has led to the artificial tailoring of crystals to our needs. Apart from various crystal growth techniques developed over the past century, treatment of materials to control or modify crystallinity is of great importance. Crystallization allows impurities to be removed, as in sugar refining; crystal size and shape can be controlled as in ice-cream, steel and salt production (fig 115).

INTERNALLY USEFUL

The inner and outer structures of crystals, with their perfection, their defects and their impurities, are exploited in surprising ways and in familiar surroundings.

Research into crystal structure gives you 'engineered' materials such as light-emitting diodes – LED's, single crystal turbine blades, transistors, imitation diamonds, tape recorder heads, tuneable solid-state lasers and gas igniters in your cooker or boiler.

You use people's knowledge and control of the ways that crystals behave whenever you cross bridges, enjoy sweets, watch TV, travel by train, car or plane, paint the front door, hold a cup, ring home, drill for oil and look at the time.

Atomic patterns in *piezoelectric* crystals vibrate to time (in quartz watches) and set up an electric charge when squeezed (in gas igniters). The quartz crystal slice (fig 120), with gold-coated 'prongs', vibrates 32 768 times each second to control watch timing.

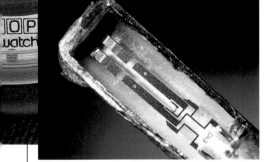

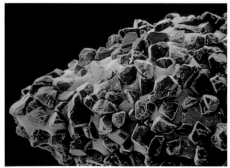

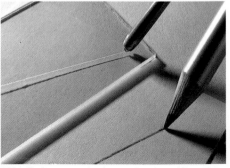

120 Quartz watch and its quartz crystal.

121 Diamond abrasive grit in your dentist's drill.

122 Carbon crystals: diamond and graphite points.

123 Clay's plastic quality comes from its crystal properties.

Research into the nature of minerals and artificial crystals has led to increasingly detailed studies of how crystals behave and how their properties can be utilized for specific purposes. For instance, natural and artificial 'crystal sieves' (fig 125) are used to separate gases and chemicals – as in water softening – and as catalysts. Their crystal structures can now be tailored to suit various special uses. Control of the degree of crystallization and crystal size is achieved by 'seeding' and by the use of colloidal particles in gels such as gelatine. 'Glass ceramic' oven ware is crystal-seeded; processed foods and sweets have crystal control for suitable texture. Growth of crystals which affect setting and impart rigidity is vital for the use of cement (fig 126) in concrete constructions. The 'flaky' pattern of the clay mineral, kaolinite, in china clay (fig 123) imparts plasticity for making your vase – as well as insulators and furnace linings. Diamond's sturdy structure makes it irreplaceable for grinding (fig 121), shaping and cutting.

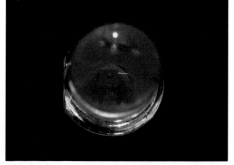

124 Light-emitting diodes (LED's) are crystalline.

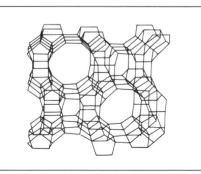

125 Molecule-sized holes in zeolite crystal sieve.

126 Crystals in cement.

IMPERFECTION RULES

Imperfect crystals keep the world going round. Slippage along crystal defects allows the Earth's rocks to move, glaciers to flow. Defects help form gems, rust and photos.

When metals are distorted, *dislocations* (fig 127) sweep through them, allowing the metal to 'bend'. However, if crystalline grains of carbide 'impurity' are present in the structure (fig 134), they hold up the dislocations and stop the metal from bending; they toughen and harden it.

Gems reveal beautiful effects of imperfection (fig 128) in reflections from minute, crystal-controlled mineral needles and layers. Twinning (fig 130) and dislocations allow solids to 'flow' slowly.

Crystal defects also produce colour; and without them transistors would not work, photography would be impossible and rusting could not take place. Lasers, LED's and transistors work with deliberately doped patterns of crystal impurities.

128 Star and cat's eye gems and moonstone.

129 Colour from salt crystal defects.

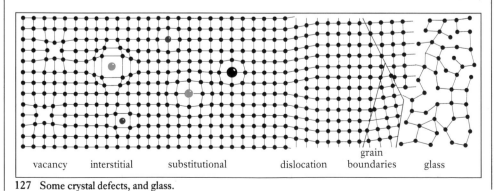

| vacancy | interstitial | substitutional | dislocation | grain boundaries | glass |

127 Some crystal defects, and glass.

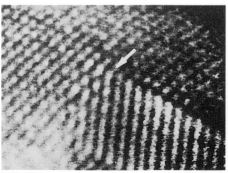

130 Twinning of atomic pattern (× 13 million).

131 Colour from crystal intergrowths, labradorite feldspar.

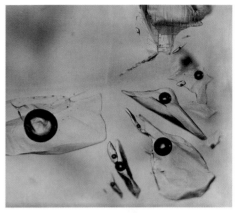

132 Fluid inside a fluorite crystal.

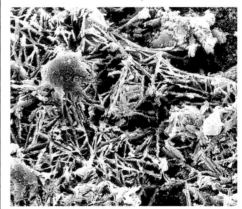

133 Close-up of car rusting.

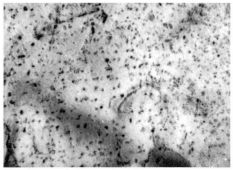

134 Tiny carbide crystals make steel hard.

CRYSTAL LORE

Crystals have for long been associated with magic and medicine, luck and direction. The 'magic' of crystals in science and adventure fiction has sometimes come true.

The crystal ball is seen as the epitome of crystal power and mystery. However, your fragile future may lie in glass rather than in a real crystal! Crystal glass is not crystalline. This real crystal ball is a rare calcite specimen and gives a double image – for alternative futures, perhaps?

Superstition has influenced people's knowledge of minerals since ancient times. The talisman or charm for protection and enhancement is often a mineral with special beauty, shape, colour or hardness.

135   A real crystal-ball: what does it tell you?

136   Islamic carnelian talisman seals, Iran, 17th–18th c.

Crystals in space fiction have intrigued readers for generations and still feature in modern space dramas. Now we can see the crystals in real Moon rock (fig. 137).

137 Crystalline rock from the Moon.

138 Sci-fi crystal comic power!

CRYSTAL TECHNOLOGY GALLERY

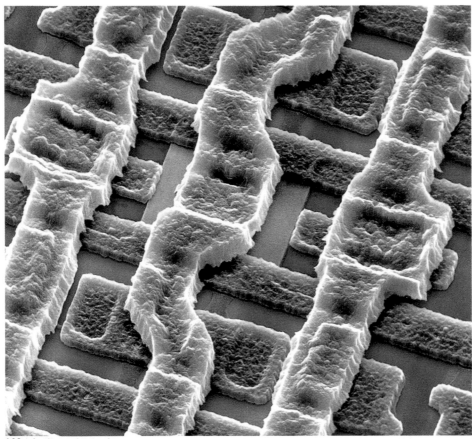

139 A silicon chip in close-up.

140 Space probe window of diamond.

141 Crystals grown for infra-red sensors.

142 Laser crystal and light.

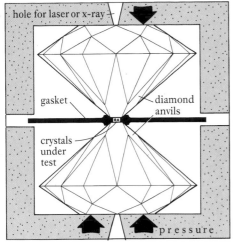

143 Diamond strength and transparency is used for high pressure crystal research.

144 Scalpel uses diamond's hardness.

145 A 'crystal set', 1923. A copper coil ('cat's whisker') on galena crystal.

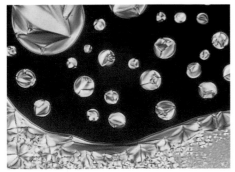

146 Liquid crystals growing.

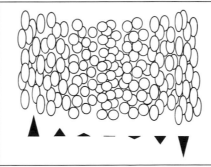

147 Liquid crystal structure.

148 Liquid crystal display.

CRYSTAL WONDER GALLERY

149 A huge amethyst geode.

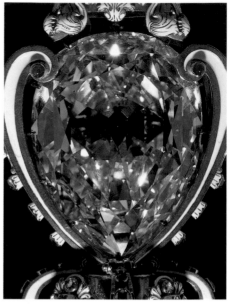

150 Cullinan I lifesize – the largest cut diamond, it weighs 106g; it is in the Tower of London.

151 An emerald crystal as it was found.

152 A diamond crystal as it was found.

153 Feldspar crystal shape replaced by tinstone: a natural trick crystal, or *pseudomorph*.

155 Galena crystal under the microscope (× 500).

154 Crystals in hand specimen (× 0.75): cubes of galena with purple fluorite, from Illinois, U.S.A.

FACTS & FIGURES

CRYSTAL & CRYSTALLOGRAPHY

The word *crystal* is from Greek, *κρύσταλλος, krystallos*, from *kryos*: icy cold. Pliny the Elder, Book 37, wrote, 'a violently contracting coldness forms the rock crystal in the same way as ice.' That rock crystal, clear quartz, is made from ice so deeply frozen that it will not thaw was an idea that existed in China, Japan and Alaska, and people believed this into the 17th century. Robert Boyle demonstrated that quartz is far too dense to be a form of ice. Solinus had reached a similar conclusion in the 3rd century BC.

Late 1660's: crystallography, the study of crystals, started in earnest. The crystal systems (see page 16) were established in the latter half of the 17th century after crystals had been studied in terms of external shape and the stacking of spheres.

1665: Robert Hooke, observing stacked musket balls, reasoned that alum crystal shape could be built up from minute spheres.

1669: Niels Steensen (Steno) established that the angle between equivalent faces on different quartz crystals is always the same, regardless of growth distortions.

1690: Christian Huygens, in his *Treatise on Optics*, explained the structure of Iceland spar (calcite – p. 41) in terms of a 3-D pattern of 'squashed spheres'.

Late in the 18th century crystal structure was looked at in terms of minute 'building block' units.

1781: René Haüy accidentally broke a calcite crystal, and its orderly cleavage angles suggested a regular stacking of elementary building blocks. His crystal structure theory, published in 1784, was a great step in the science of crystallography.

Around 1780: Romé de l'Isle measured many crystal face angles, and the *Law of Constancy of Angle* was established: *in all crystals of the same substance, the angles between corresponding faces have a constant value.*

An early wooden model octahedron showing small cube units as 'building blocks'.

1830: J F C Hessel derived the 32 different 'point groups' of 3-D symmetry, also known as the crystal *classes* (see page 16).

1848: Auguste Bravais found that there are just fourteen different types of arrangement of regularly-repeating points in three dimensions: the 'Bravais lattices' (see page 5).

1885: Fedorov established the existence of 230 different sorts of 3-D patterns. These 'space groups' represent the different ways that groups of atoms 'clothe' the fourteen 'Bravais lattices' (see page 15).

1912: X-rays took the study of crystals deep into their atomic structure. Max von Laue reasoned that atomic patterns of crystals should help him establish the wavelength of X-rays. W Friedrich and P Knipping aimed an X-ray beam at copper sulphate crystals and obtained successful diffraction patterns.

1913: W H and W L Bragg turned the experiment around: using the newly-found wavelength value, they used X-rays to determine the structures of various crystals using X-ray diffraction. The first crystals to have their structures analysed were potassium chloride and sodium chloride.

Crystal structures were now seen to be built up in repeating units of atomic pattern rather than from solid shapes.

Clay model of pyrite (left) made to illustrate the work of Romé de l'Isle for his *Crystallographie*, 1783.

Wooden model of calcite (right) made by Count de Bournon, in the Greville Collection obtained by the Natural History Museum in 1810.

SYSTEMS & SYMMETRY: CLASSES

The table shows the eleven most common crystal classes (see page 16) found in minerals.

There are 32 classes in all. Some are rare. A typical mineral is named for each class. Classes are also called 'symmetry groups', or 'point groups'. The number of each kind of symmetry possessed by each class (see pages 14 and 15) is listed:

SYSTEM NAME:	Kinds of symmetry –					
	Axes: (×-fold)				Planes:	Centre:
	2	3	4	6		
CUBIC						
diamond	6	4	3	–	9	yes
pyrite	3	4	–	–	3	yes
tetrahedrite	3	4	–	–	6	no
TETRAGONAL						
zircon	4	–	1	–	5	yes
TRIGONAL						
quartz	3	1	–	–	–	no
calcite	3	1	–	–	3	yes
tourmaline	–	1	–	–	3	no
HEXAGONAL						
beryl	6	–	–	1	7	yes
ORTHORHOMBIC						
topaz	3	–	–	–	3	yes
MONOCLINIC						
orthoclase feldspar	1	–	–	–	1	yes
TRICLINIC						
plagioclase feldspar	–	–	–	–	–	yes

Bronze pivoted models to show twinning in gypsum (right) and spinel; made before 1811.

CRYSTALS IN SYSTEMS

CUBIC galena, pyrite, garnet, diamond, spinel, halite (common salt), fluorite, zinc blende (sphalerite), alum, magnetite, tetrahedrite, cuprite, pitchblende, copper, gold, silver, iron, lead, platinum, silver bromide, quicklime (calcium oxide).
TETRAGONAL cassiterite, zircon, wulfenite, scheelite, scapolite, idocrase, chalcopyrite, rutile, urea, potassium hydrogen phosphate.
TRIGONAL tourmaline, corundum (ruby and sapphire), quartz, calcite (incl. coral, stalagmite), dolomite, magnesite, hematite, dioptase, cinnabar, chile saltpetre (sodium nitrate), carborundum (silicon carbide).
HEXAGONAL ice, graphite, apatite, beryl (incl. emerald and aquamarine), nepheline, zinc, iodoform, silver iodide, menthol.
ORTHORHOMBIC baryte, topaz, olivine (peridot), nitre (saltpetre, potassium nitrate), staurolite, chrysoberyl, epsomite (epsom salts), aragonite, alpha-sulphur, cordierite (iolite), zoisite (incl. tanzanite), stibnite, anhydrite, marcasite, natrolite, rochelle salt, codeine, strychnine, atropine, oxalic acid, citric acid.
MONOCLINIC orthoclase feldspar (incl. moonstone), gypsum (selenite), mica, jade (jadeite and nephrite), steatite (soapstone, talc), asbestos, malachite, chlorite, epidote, amphibole (incl. hornblende), pyroxene (incl. augite), serpentine, clay minerals (including kaolinite), spodumene, borax, beta-sulphur, sucrose (and other sugars), tartaric acid, washing soda, sodium bicarbonate (in baking powder), naphthalene, glauber's salt.
TRICLINIC turquoise, plagioclase feldspar (incl. labradorite and sunstone), rhodonite, axinite, kyanite, copper sulphate.

Part of a collection of 24 cut glass models of the crystal forms of precious stones, in natural colours; M. de Struve 1825.

THE WORLD'S LARGEST CRYSTALS

THE WORLD'S LARGEST AUTHENTICATED CRYSTAL:
A beryl crystal, Malagasy Republic.
18 metres long, 3.5 metres in diameter, weight: about 380 tonnes.

Other large mineral crystals:

mineral		size (m)	weight (kg)
fluorite	New Mexico	2.13	16 000
calcite	Iceland	$7 \times 7 \times 2$	254 000
garnet	Norway	2.3	37 500
topaz	Mozambique	0.91	2677
mica	India	4.57×3.05	77 000

● A quarry in Colorado, USA, might have been in a single crystal of microcline feldspar.
Its size would have been $49.38 \times 35.97 \times 13.72$ metres, with an estimated weight of about 16 000 tonnes.
● The largest known crystal from space is a single iron-nickel alloy crystal in a meteorite found in USSR.
It measures $920 \times 540 \times 230$ mm, weighs 303 kg.
● The largest diamond crystal known was the Cullinan found in 1905 in South Africa. It weighed 3106 carats (about 621 g).

THE LARGEST FACETED CRYSTALS
● The largest cut diamond is the Cullinan I, 530.20 carats, in the British Crown Jewels (see page 57).
● The heaviest faceted gem is a yellow topaz, at 22 898 carats (over four and a half kilograms).
● The largest cut gem by volume is a citrine (yellow quartz), at $255 \times 141 \times 100$ mm: 19 548 carats. (5 carats = one gram).

Late 19th C. wooden model hexoctahedron, one of a set of over 700 accurate pearwood models made by the firm of Krantz in Bonn.

OTHER RECORDS
● The longest known stalactite (from the cave roof) is 7 m long (The Poll Cave, County Clare, Ireland).
● The longest stalagmite (from the cave floor) is 29 m long (Aven Armand Cave, France).
● The longest complete column is 39 m long (Nine Dragons Cave, China).
● The smallest brilliant cut diamond is 0.22 mm in diameter, weighing 0.0012 carats (0.00024 g).

INTERATOMIC DISTANCES
in nanometres (nm)
(one nanometre = a millionth of a millimetre).

In *quartz* and other silicate minerals (see pages 8–11): silicon to oxygen distance ranges from 0.152 to 0.160 nm.

In *ice* the distance between oxygen atoms is 0.26 to 0.27 nm (a very 'open' structure).

In *diamond* the carbon atoms are 0.154 nm apart;
In *graphite* they are 0.145 nm apart (in parallel planes) and 0.335 nm apart (across the planes; see pages 28, 29).

The photograph on page 3 is a picture of the mineral epidote, scanned under a 'tunneling electron microscope'. It shows up individual atoms at a magnification of almost ten million, revealing the crystal's monoclinic pattern in a direction parallel to its reference b-axis (also parallel to its two-fold symmetry axis).

Rhombohedron in glass sheet, with thread axes; from a set made to order by Dr Krantz in Bonn, 1896–1901.

INDEX

Designer: David Robinson
Artist: Paul Allingham – Panda Art
Photographers: Frank Greenaway, Harry
    Taylor, Natural History Museum
Typesetting by Cambridge Photosetting
    Services
Printing & Binding in Singapore by
    Times Offset (S) Pte Ltd.

Other picture sources:
Figs 1, 16, 18, 37, 38, 47, 49, 52, 53, page
    26, 75, 77, 78, 80, 116, 121, 130, 133, 139,
    142, 146, 148 Science Photo Library
Figs 6, 86 Cassandra Goad
Figs 17, 51 British Geological Survey
Figs 19, 22 (inset), 24, 25, 27, 36, 40, 43–
    46, 56, 60, 65, 68, 71, page 38, 93, 95,
    98, 102, 128, 131, 137, 151, 152
    Geological Museum
Figs 22, 69, 83 Katia and Maurice Krafft
Fig 35 Spink & Son Ltd
Fig 36 (inset) Eric Robinson
Fig 39 Christine Woodward
Figs 42, 110 Ian F Mercer
Fig 48 E Alan Jobbins
Fig 50 Kodak Limited, Research
    Division/Journal of Photographic
    Science
Front cover: figs 54, 55, 59, 61, 63, 69
    (inset), 89, page 39, 94, 101, 111, 129,
    132, 153–155; back cover: Department
    of Mineralogy, Natural History
    Museum, London.
Fig 73 Elisabeth Bauser
Fig 88 Roy Huddlestone
Fig 108 R F Symes
Fig 115 ICI Chemicals and Polymers
Figs 117, 118, 141 Crystalox Ltd
Fig 120 Louis Newmark plc/Swatch
Fig 123 (inset) English China Clays plc
Fig 123 Josiah Wedgwood & Sons Ltd

Fig 126 Karen L Scrivener
Fig 134 Highveld Steel & Vanadium
    Corporation Ltd
Fig 136 Michael and Henrietta Spink
Fig 138 Jean Giraud – Moebius/H Bari
Figs 140, 144 De Beers Consolidated
    Mines Ltd
Fig 145 The Vintage Wireless Museum,
    West Dulwich
Fig 149 Hubert Bari
Fig 150 photograph taken by gracious
    permission of Her Majesty The
    Queen; Crown copyright reserved
Page 3 G Cressey
Page 8 NASA/Geological Museum
Page 27 P Krishna

Assistance with specimens for
photography; script checking and
discussion: The Department of
Mineralogy, Natural History Museum,
London; with particular help from
Dr R F Symes

We are grateful to David Kent – for
assistance with figs 6, 86 and 112, and to
Brian Lloyd for assistance with fig 109.

© British Museum (Natural History) 1990

All rights reserved.

**Library of Congress Cataloging-in-
Publication Data**
Mercer, Ian, *1944–*
    Crystals.
    1. Crystals. I. Title
QD921.M463 1990   548   89-26799

ISBN 0-674-17914-5

Reprint 10 9 8 7 6 5 4 3 2 1

Metric conversions

gram	= 0.035 ounce
tonne	= 1.1 ton
metre	= 39.37 inches
kilometre	= 0.62 mile
0°C	= 32°F